WHAT ARE THE FACTS?

Collecting Information

Barbara A. Somervill

The Rosen Publishing Group's
PowerKids Press™
New York

For Alexandra Renee

Published in 2007 by The Rosen Publishing Group, Inc.
29 East 21st Street, New York, NY 10010

First Edition

Editor: Joanne Randolph
Book Design: Elana Davidian
Layout Design: Julio Gil

Photo Credits: Cover, p. 19 © JLP/Jose Luis Pelaez/zefa/Corbis; p. 4 © Michael Prince/Corbis; p. 7 © Ed Bock/Corbis; p. 8 © Jim Craigmyle/Corbis; p. 12 © Ted Horowitz/Corbis; p. 15 © Ryan McVay/Getty Images.

Library of Congress Cataloging-in-Publication Data

Somervill, Barbara A.
  What are the facts? : collecting information / Barbara A. Somervill.— 1st ed.
      p. cm. — (Think like a scientist)
  Includes bibliographical references and index.
  ISBN 1-4042-3484-5 (lib. bdg.) — ISBN 1-4042-2193-X (pbk.)
  1. Science—Methodology—Juvenile literature. 2. Experimental design—Juvenile literature. 3. Research—Juvenile literature. 4. Observation (Scientific method)—Juvenile literature. I. Title. II. Series.
  Q175.2.S663 2007
  501—dc22
                                                2005035722

Manufactured in the United States of America

Contents

The Scientific Method 5

Research and Keywords 6

Evaluating Resources 9

Measurement Tools 10

Data and Observations 13

The Project Log 14

Daily Entries 17

Comparing Your Results 18

Making Tables, Graphs, and Diagrams 21

What Did You Learn? 22

Glossary 23

Index 23

Web Sites 24

These two boys are working on an experiment. As they follow the scientific method, they will collect information about what happens when they mix these liquids together.

The Scientific Method

Scientists spend most of their time collecting **information**. They follow the **scientific method** to collect this information. They start with a subject, come up with a **hypothesis**, and run experiments to prove their ideas are correct.

Research, or careful study, is one form of fact gathering. Running experiments, making observations, and recording activities are other ways to collect information. You begin collecting information from the moment you start thinking about a science **project**.

You must keep careful records of everything that you do, observe, or measure. Some information will prove your ideas. Some information will not.

Research and Keywords

Information gathering starts with research. Let's say that you see a bridge that carries tons of weight every day. You might wonder how the bridge can hold so much weight. Bridges are long and narrow, and they are often made with thin beams of steel. You decide to find out whether building **designs** have to do with the strength of a **structure**. To start your research, list keywords related to your subject.

For this project keywords might include "building structures" and "building **materials**." You might do an Internet search. Look at the list of books in the back of **engineering** books to help you find information.

You must keep looking until you find the information you need. Record your findings by taking notes or making copies.

Using the Internet at the library is one way to collect information on your project's subject. Do not forget to check the library's catalog, too.

A builder, such as this man, knows that with the right design he can use materials that weigh little compared to the weight they will hold. A builder collects information that will help him do his job well.

Evaluating Resources

Not every book or magazine article will have the information you need. If you look at 20 **sources**, you may find five that are helpful. The best sources have reliable, or trustworthy, information.

Reliable information comes from many sources. Scientific groups, such as NASA and the World Wildlife Federation, produce useful books and Web site articles. Newspapers and magazines are also good sources. They usually have researchers check the information they print.

If you are finding out the strength of lightweight structures, you might also ask an engineer or builder questions. These people know that the way a structure is built can give it strength.

Measurement Tools

Before you start work on any experiment, you need to figure out how you will measure the results. Height, weight, volume, and time are common experiment measures. You can use tools such as rulers, scales, or **thermometers** to find these measurements.

For your experiment you might decide to build a **pyramid**, a **helix**-shaped solid, a **cube**, and a prism using bamboo sticks and modeling clay. To test how much weight each model can hold, place a paper plate on each structure. Slowly add pennies until the structure falls down. Weigh the pennies and the plate used for each structure with a metric kitchen scale. If you have no scale, use the number of pennies as your measurement.

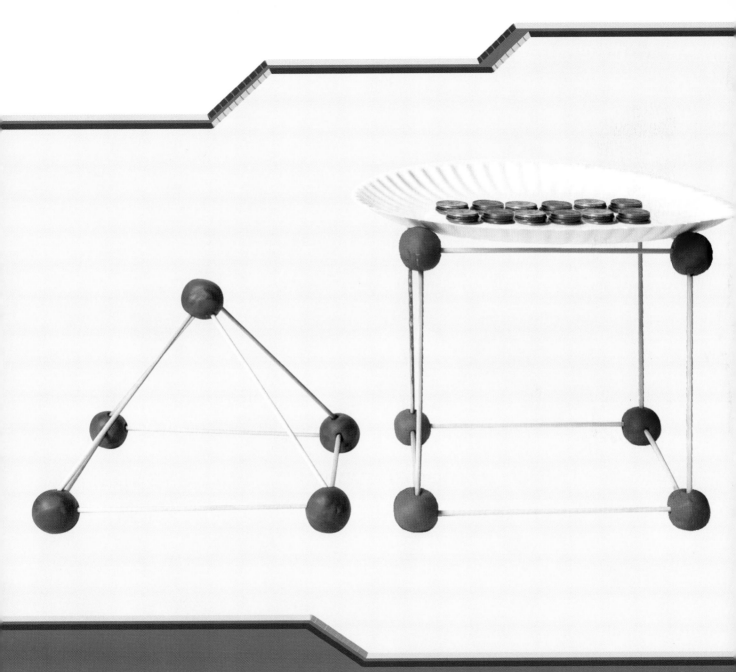

Here we see two of the structures built with bamboo sticks and clay. You cannot just say that one structure holds more weight than another. You must give information that backs your statement up.

This boy is building a model of a double helix to help him collect information about DNA. DNA is the part of your cells that decides what you look like. You look like your parents because you share their DNA.

Data and Observations

Your collected information includes data and observations. Data are facts gained from an experiment. Data include your measured results. Observations are events that you see, hear, feel, smell, or taste. Record data and observations as you work.

In an experiment you should have both data and observations. You built four structures for your experiment. Your data should list the size and number of bamboo sticks and clay connectors you used. The amount of weight each structure holds is also data.

You might find that adding weight to the pyramid structure is hard. You then see that the helix-shaped structure is the only one with sticks crossing the structure's center. You also find that the helix-shaped structure holds the most weight. Those are observations.

The Project Log

Start each project by creating a project log. This should be a bound notebook. Do not use a spiral-bound or loose-leaf notebook because pages might get lost.

You must number each page before you start to work. Do not worry about making mistakes or messy smudges. What counts in a project journal is not neatness but completeness. Just make a note on the edge of the page saying that material on this page has a mistake.

Put colored tabs on pages to show where actions, materials, and measurements are listed. For example, choose one color for actions, another for materials, and a third for measurements. Colored tabs will make finding information for your report easy.

A notebook is the perfect place to keep track of all your information. Take careful notes and write down all the data and observations from your experiment. This way others can make sense of your results.

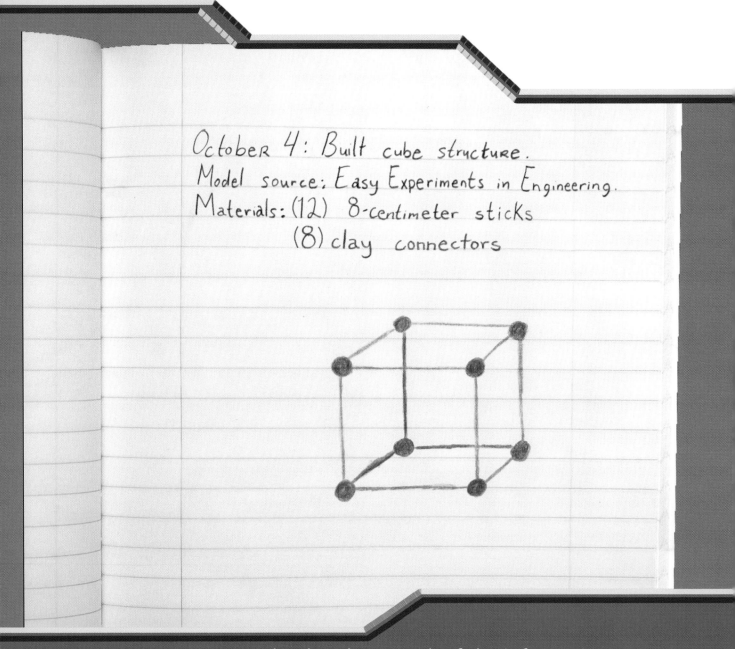

October 4: Built cube structure.
Model source: Easy Experiments in Engineering.
Materials: (12) 8-centimeter sticks
(8) clay connectors

You can use your notebook to keep track of the information you collect. This page shows what one student did on October 4.

Daily Entries

Experiments take time, and your journal records the events that take place. You will want to make entries daily, as you would in a diary.

Write down the date first. Tell what actions you took on that date. If you have a certain source you used for this action, include the name of that source. For your experiment record, you want to include everything you did, all the materials you used, and the proof you have.

As you work record everything about your experiment in your journal. You must write down activities, observations, and measurements. Write the date for every entry, then add your information. You should even date pictures.

Comparing Your Results

Do not forget that other people's experiments can provide you with information. These experiments might appear in a book, on the Internet, or in your classroom.

You can learn from other people's work. They may have followed different steps in their experiments. They may have used different materials than you did. You need to figure out if using any of their ideas would improve your experiment.

Suppose two other students worked on an experiment with the same hypothesis yours had. All three of you guess that the shape of a structure adds to its strength. One built structures with uncooked spaghetti and **marshmallows**. Another used straws and foam balls. You all found that helix-shaped models held the most weight. Your classmates' experiments back up your findings.

These students are working together on a science experiment. They will share their results with the rest of the class. By comparing information they can find out whether they had the correct hypothesis.

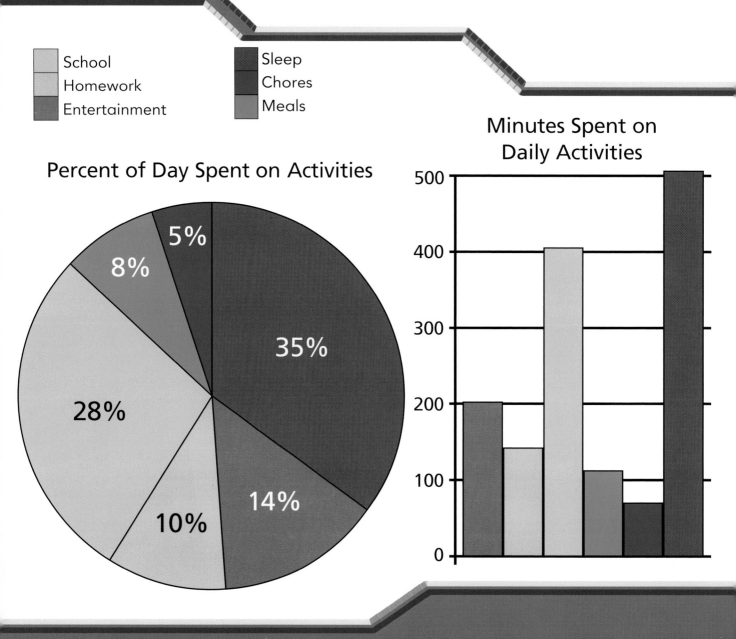

School
Homework
Entertainment
Sleep
Chores
Meals

Percent of Day Spent on Activities

5%

8%

35%

28%

14%

10%

Minutes Spent on Daily Activities

500

400

300

200

100

0

One way to collect information is to make charts, graphs, and diagrams. Here we see a pie chart and a graph. Graphs and charts can be much easier to understand than a list of numbers.

Making Tables, Graphs, and Diagrams

In class you use graphic **organizers** to help sort, record, and understand information. You can use the same tools, such as tables, graphs, charts, diagrams, and drawings, in your experiment.

A table with **columns** and rows is useful for recording measurements. Each column could show data about a different model. List the data collected for each measurement. A graph can be created from the data in the table. A graph offers a quick way to compare results for all structures.

Create diagrams and drawings to help describe either actions or results. A diagram could show what a model looked like. A step-by-step drawing gives a picture of how each model was put together. Graphic tools will help people better understand your work.

What Did You Learn?

It is not enough to have a good idea for a science project. You must back up your ideas with facts, and those facts need to be written down. All your plans, ideas, research, and results make up your project information. Collecting that information makes up most of the work you will do.

Your project log is like an experiment diary or journal. It will remind you of everything you did, observed, and measured. There is simply no way to keep all that information in your head.

The time spent keeping good records pays off in the end. When it is time to make your presentation, creating a display and writing a report will be quick and easy. After all you have been collecting information from the beginning.

Glossary

columns (KAH-lumz) Rows that go up and down, instead of side to side.

cube (KYOOB) A shape with six square sides.

designs (dih-ZYNZ) The plans or the forms of things.

engineering (en-juh-NEER-ing) The science of planning and building such things as engines, machines, bridges, and buildings.

helix (HEE-liks) Something that has a spiral shape.

hypothesis (hy-PAH-theh-ses) A possible answer to a problem.

information (in-fer-MAY-shun) Knowledge or facts.

marshmallows (MARSH-meh-lohz) Spongelike foods made from whipped sugar.

materials (muh-TEER-ee-ulz) What something is made of.

organizers (OR-guh-ny-zerz) Tools to make facts neat and easy to understand.

project (PRAH-jekt) A special job that someone does.

pyramid (PEER-uh-mid) A shape that has a square base and triangular sides that meet at the top.

scientific method (sy-un-TIH-fik MEH-thud) The system of running experiments in science.

sources (SORS-ez) Things that give facts or knowledge.

structure (STRUK-churz) Form.

thermometers (ther-MAH-meh-terz) Tools used to measure how hot or cold something is.

Index

D
data, 13

E
experiment(s), 5, 10, 13, 17–18

G
graphic organizers, 21

H
hypothesis, 5, 18

I
Internet, 6, 18

M
measurements, 17, 21

N
NASA, 9

O
observations, 5, 13, 17

P
project log, 14, 22

R
records, 5, 17
research, 5–6, 22

researchers, 9
results, 10, 13, 22

S
scientific method, 5
source(s), 9

W
World Wildlife Federation, 9

Web Sites

Due to the changing nature of Internet links, PowerKids Press has developed an online list of Web sites related to the subject of this book. This site is updated regularly. Please use this link to access the list:
www.powerkidslinks.com/usi/collinfo/